Agile During a Pandemic

By Christina Simpson

Published by Christina Simpson © 2020 Christina Simpson, Minneapolis, MN as of March 03, 2021 Content is based on personal work experience and common knowledge of the author, except where clear external materials are referenced. The author and publisher's intent are only to, in good will, share information with others on the Agile Journey. The author and publisher are not responsible for the actions of readers.

All rights are reserved. No part of this book may be reproduced for sale or modified in any form, including photocopying, recording, or by any information storage and retrieval system, without permission in writing from the publisher.

Extensive effort has been made to acknowledge and properly cite any external materials referenced in this work. If copyrighted materials have inadvertently been used in this work without credit and proper citation. Please contact the author via written or email correspondence so that future digital and print copies of this work can be corrected in compliance.

ASIN : B08RCD7R3F

Publication date : March 3, 2021

Language : English

Simultaneous device usage : Unlimited

Text-to-Speech : Enabled

Enhanced typesetting : Not Enabled

X-Ray : Not Enabled

Word Wise : Not Enabled

Lending : Enabled

ISBN: 9798587051874

Imprint: Independently published

Content

Introduction..5

Chapter 1: Big Picture, What Is Agile?..9

    Agile Manifesto

Chapter 2: Impact of COVID-19 on Organizations......................................13

    Agile as a Solution

Chapter 3: Impact of COVID-19 and George Floyd Civil Unrest on People..................24

    Leveraging the Agile Culture

Chapter 4: My Experience Being Hired During the Pandemic......................38

    Agile Coach in Healthcare Tech

Chapter 5: Agile on the Job..47

    Applying Agile during the pandemic

        Work Environment..49

        Building Relationships Remotely..54

        Face to Face Interactions..55

        Video Engagement..55

        One on Ones..58

        Team Building...59

        Satisfying Customers and Continuous Delivery..........................62

        High Level Metrics do Help..65

        Adjusting to Change..69

        Business and Development Work Together................................70

        Retrospectives..72

        Agile Mishaps...74

Chapter 6: Conclusion...80

Introduction

Between the COVID-19 pandemic and Minneapolis George Floyd civil unrest, Agile is proving to be a survival tactic through 2020. Non-Agile offices around the world are suffering from a loss of profit and employees. Agile competitors are amplifying Agile methods as quickly and widely as possible.

As the Boomer generation exits the workforce for retirement and millennials increase their presence, new ways of doing work which better meets the needs of both millennial workers and consumers are emerging. Having these two generations interacting in so many ways, brings many different expectations and challenges to the marketplace.

Millennial voices are growing louder. Millennial workers want positive work cultures, work life balance, flexibility and autonomy and will leave one organization for another, if these attributes are not met. As customers they are showing strong preference for immediate delivery of products or services. The Agile train has picked up speed and vitality as a result of 2020.

The pandemic has created a new level of chaos and the need to be adaptive. What is known about the virus, how it spreads and how to protect yourself seems to change monthly, if not weekly. Businesses need to be able to shift quickly and Agile is perfect for this.

It has become more vital than ever before to work effectively while adapting to changing circumstances. The pandemic is regularly changing what is considered essential and what is not, how and what products get to the market, etc. Organizations are experiencing workforce instability with illness and quarantine. These pandemic effects are unpredictable. Who is available and who is not changes rapidly and may last weeks to months. These impacts can be huge, in some cases entire departments may be out sick or quarantined at one time. Agile has components that can be used as a heavy-hitting antidote to these dynamic landscapes. Organizational longevity depends on the ability to be adaptive during this time. The name of the game is becoming comfortable with the unknown.

Agile During a Pandemic will walk you through Agile basics. What is Agile at its core? How can you live Agile? Christina tells her story as a Minneapolis based, healthcare Agile Coach during the COVID-19 pandemic and George Floyd events.

Whether you're a well-seasoned Agilist, or a newcomer to the Agile world, *Agile During a Pandemic* offers something for everyone. The opinions expressed in this book are the author's own based on personal experience.

I want to start off by welcoming everyone and introducing myself. I'm Christina, an Agile Coach. I joined a small healthcare tech start-up during the tumultuous and unprecedented time of the COVID-19 pandemic.

In March 2020, COVID-19 in the United States was just a few months past its first confirmed positive COVID-19 case. We got a glimpse of what was coming from watching the news and seeing other countries all around the world as they went into a spiral, rapidly losing control. The American people knew something unknown and scary was on the rise.

Chapter 1

Big Picture, What Is Agile?

Let's talk a little bit about what Agile means. Rather than giving my interpretation, I want to provide you with the original foundation of Agile directly from the source, so we're starting on the same page for further discussions.

The following Agile Manifesto principles and values are downloaded and copied verbatim from https://agilemanifesto.org on February 2, 2021.

"Manifesto for Agile Software Development"

"We are uncovering better ways of developing software by doing it and helping others do it. Through this work we have come to value:"

*"Individuals and interactions over processes and tools
Working software over comprehensive documentation
Customer collaboration over contract negotiation
Responding to change over following a plan*

That is, while there is value in the items on the right, we value the items on the left more."

"Principles behind the Agile Manifesto"

"We follow these principles:"

"Our highest priority is to satisfy the customer through early and continuous delivery of valuable software."

"Welcome changing requirements, even late in development. Agile processes harness change for the customer's competitive advantage."

"Deliver working software frequently, from a couple of weeks to a couple of months, with a preference to the shorter timescale."

"Business people and developers must work together daily throughout the project."

"Build projects around motivated individuals. Give them the environment and support they need, and trust them to get the job done."

"The most efficient and effective method of conveying information to and within a development team is face-to-face conversation."

"Working software is the primary measure of progress."

"Agile processes promote sustainable development. The sponsors, developers, and users should be able to maintain a constant pace indefinitely."

"Continuous attention to technical excellence and good design enhances agility."

"Simplicity--the art of maximizing the amount of work not done--is essential."

"The best architectures, requirements, and designs emerge from self-organizing teams."

"At regular intervals, the team reflects on how to become more effective, then tunes and adjusts its behavior accordingly."

*Kent Beck
Mike Beedle
Arie van Bennekum
Alistair Cockburn
Ward Cunningham
Martin Fowler*

*James Grenning
Jim Highsmith
Andrew Hunt
Ron Jeffries
Jon Kern
Brian Marick*

*Robert C. Martin
Steve Mellor
Ken Schwaber
Jeff Sutherland
Dave Thomas*

© 2001, the above authors

this declaration may be freely copied in any form,

but only in its entirety through this notice.

The Agile adoption process is about adopting a lifestyle, culture, mindset founded on very simple and straightforward principles and values. Agile as a "project management method" is not going to work unless the real mentality and culture of Agile is adopted.

Chapter 2

Impact of COVID-19 on Organizations

A global pandemic has claimed the lives of over two million people so far worldwide and nearly a half a million here in the United States. 2020 is a year that demands flexibility, it requires the ability to embrace the unknown, and it requires the ability to change directions quickly with minimal loss.

It requires human interaction to be the priority over the process. It demands taking things one deliverable step at a time with full transparency. There is a greater need than ever before for quick and continuous feedback loops. The need to continuously inspect and adapt is proving critical to the survival of every company.

These are unprecedented times with a completely unpredictable, uncertain, unfathomable haze covering every area of life. The COVID-19 virus is a source of legitimate fear. The rest of the uncertainty and instability of venturing in uncharted pandemic waters (for an Agile Coach) constitutes an epic once in a lifetime opportunity to put all Agile tools to use.

Easy to talk -the -talk about embracing change when the conditions are normal, but are you up for the challenge even when the risks are high and conditions overwhelming?

Many companies here in Minneapolis, MN have switched to remote options for 'non-essential' workers and any essential workers who have the capabilities in their role to work remotely. My current organization has been a remarkable example of how to do this with grace and Agility.

I'll present to you a picture of how things needed to change quickly and drastically. This drastic change is nothing to be taken lightly. Many organizations sink or swim based on their ability to adapt Agile.

Picture a strategic plan finalized in Q4, 2019 for 2020 based on alignment with customer needs, careful analysis of priorities, customer feedback, and return of investment. Sounds normal, right?

Now picture those freshly finalized plans being thrown out the window of a truck, going down the highway at 70 miles an hour. For many companies in 2020, all of their plans were lost overnight as hundreds of thousands of people lost their jobs.

Many organizations could not survive the initial government-mandated lockdown. Hundreds of thousands of people fell ill through the early stages of this pandemic. Turn on the news channel and see makeshift morgues being set up in major cities. Get two weeks' worth of food, the city is locking down in 24 hours.

Suddenly companies were scrambling in a complete panic to figure out two things.

Number one, how to help? Many organizations not mandated to close found ways to slow the spread of the virus by having selected employees work remotely. Many employers also generously paid employees who were not able to come to work. This pandemic showed how truly interconnected we all are, with a spread rate doubling and tripling overnight.

Number two, how to survive?!!! Many strategic plans in 2020 were completely outdated and irrelevant about one month into the new year. Leaders everywhere hit the drawing board asking questions like what parts of the business can be continued and how will it be done?

Restaurants are an example. In the beginning, they transitioned from indoor to outdoor seating then to take out orders only. Even with these adaptations a significant number of employees had to be laid off.

Doctors' offices went from in person visits to virtual visits for non-emergency issues. Hairdressers went from dying hair in a salon (providing a service) to delivering pre-mixed hair dyes with directions for their clients (providing a product). What do clients need right now and can all efforts shift in that direction quick enough?

Suddenly the abilities of organizations to get work done in a chaotic situation was front and center. Each method comes with its own pros and cons during the pandemic.

Waterfall is a more traditional approach to doing work. Work is planned with as much detail upfront as possible. Target dates and deliverables are assigned to developers. Waterfall's overall vibe is command and control.

In Waterfall the entire software development life cycle takes place, and then at the very end, the product is delivered. With a fixed budget, predetermined delivery date, product deliverables and client commitments a lot of things can go wrong. Without shorter feedback loops, a fixed plan is likely to be outdated before it is completed, resulting in delayed or unsatisfactory delivery.

During the COVID-19 lock down when products/services changed overnight, a Waterfall method with its phase gates could result in serious delays. In Waterfall it commonly takes weeks, months or years before an organization delivers a finished product. COVID-19 offers no grace period for businesses that can't immediately shift and deliver.

"Fire drill syndrome" is common in many organizations. It is the presence of little or no method at all. Fire drill is a reactionary way of completing work. It's a constant state of band-aid over band-aid to put out the immediate fire and run to the next fire.

Fire drill organizations during COVID-19 graduated to wildfire organizations. The inability to manage multiple fires during the pandemic spelled disaster for many organizations. It would not surprise me if surviving fire drill organizations turn to Agile for their business rehabilitation.

COVID-19 poses an interesting opportunity for organizations like this. There's an old saying "what doesn't kill you makes you stronger". The fire drill may have finally burned enough to make people never want to look back at old ways of doing things after the losses that occurred.

These organizations also have a rope of opportunity they can grab onto during the pandemic. Much of their workforce and customer base is equally lost about what to do next. This means people associated with these organizations may be more open and willing to try something, anything new. In normal times these same organizations may have put up more of a fight to keep things the same.

"Move fast and break things". – Mark Zuckerberg made this Facebook's main motto in its early stages. It's extremely common and yes you can get very far, very fast. A strength of this motto is that it encourages people to go with the flow and embrace unknowns.

The downside of that motto is that it can be a slippery slope risking employee burnout, high turnover and product crash. Results of this nature come from not taking the time to build out a solid ecosystem for the product to grow in. Employee turnover is expensive for companies, due to loss of mission critical knowledge. It disrupts relationships with vendors and causes delays in product delivery, due to time required to hire and train new employees.

Agile with its concern for people over process approach avoids employee burnout and turn over. Rather than delivering everything at once and waiting until the end to get feedback, Agile allows for continuous client/stakeholder evaluation and constant delivery throughout the process. With this approach, Agile avoids the crisis that inevitably comes from stiffness in planning.

During COVID-19, the ability to execute on continuous delivery has created a safety net. An Agile organization has the advantage of being able to immediately deliver something, while buying time to think through next steps for their product.

In pure Agile no one should set dates for the teams telling them, when features must be done. However, deadlines are real. There is no way around them. Working with clients and stakeholders to define the must haves versus nice to haves is a critical collaboration during these times.

Prioritizing, collaborating and strengthening the organization's product team and the operation's team will be hugely helpful in getting the desired results. Let's revisit how we're embracing change.

"Adjusting to changing requirements even late in the game." – Agile Manifesto

Priorities can shift quickly. During the COVID-19 crisis someone could reasonably ask, what could be more important? February 2021 and the Polar Vortex forced another crisis on an already stressed system causing power outages and subsequent water shortages. With millions of people without power and running water, COVD-19 vaccine distribution centers needed to quickly become water distribution centers.

In a world full of cataclysmic events, climate change, etc. the need for quick adaptation, as well as maintaining employee mental, emotional and financial security is of paramount importance for organizational survival.

Chapter 3

Impact of COVID-19 and George Floyd Civil Unrest on People

The weekend before I started my new Agile Coach job, things began to feel a little "unusual" and creepy all around town. People knew that there was some sort of lockdown coming, but no one could begin to imagine what that would be like. Around this time, people were panic buying many things, but especially panic buying.... toilet paper.

Panic stocking was happening at every store. People were stocking up to live at home the rest of their lives. I don't know precisely what people were doing with thousands of rolls of toilet paper. Others just trying to get regular weekly necessities were left with nothing when they went to the stores.

If you walked down an aisle in the store, people were so scared of sharing air space that they would turn and run the other way. Many store shelves were empty. Rather than toilet paper, there are so many other, more essential items people could have stocked up on during an apocalypse. The vitamin racks which could strengthen your immune system - those racks? Totally full. No trouble finding vitamins.

As Americans watched different parts of the world go into lockdown, it was not a surprise when the United States followed suit. For many people the pandemic was a wake-up call on interpersonal connectedness and social responsibility.

Not wearing a mask may not affect your health, but may kill someone else. Deciding to buy excessive quantities of essential items may leave someone else in need with nothing. If you come to work sick, you may risk your co-workers needing to self-quarantine for two weeks, or worse.

What is true for society is also true for an Agile team. There are no sole contributors on an Agile team. Agile teams' function as an interdependent unit. The actions and well-being of all members on the Agile team determine the team's outcome. The Agile Coach must be sensitive to the needs of team members and the impact on team output.

There are many single parents who couldn't work from home because they were deemed essential, yet their children's schools were closed. There are also many low income or rural households with no internet connection, no devices, etc., who were being forced to make significant adjustments on short notice. Sadly, this pandemic preyed on those already struggling and hit them the hardest.

Families may have lost usual methods of getting or giving support in caring for children or aging parents. Things people needed such as doctor visits, surgeries, dental and vision appointments, physical or speech therapist appointments were all canceled or inaccessible.

Typical ways of coping with stress such as working out at the gym, getting a massage, eating out, seeing a movie, visiting with friends, going to the beach, etc. were all off limits during the pandemic lock down. Self-care routines that make people feel confident about themselves such as going to the hairdresser, personal trainer, nail salon, barbershop, teeth whitening, waxing salons, eyebrows/eyelash salons all became unavailable. In the workforce, it is in this context that you are building relationships.

Prior to the pandemic work was largely just a part of many busy active lives. Now work occupies a space where the workforce also lives and teaches their children, without the clear boundaries between home, work and children.

People in this pandemic are juggling stressors without the social support network of family and friends who used to be able to help. Confidence in physical appearances and overwhelming stress may challenge the availability and willingness of team members to participate in team activities including on camera videos. On camera confidence and engagement with team members during this pandemic comes from building and having strong relationships.

As you can see, a key element of success for the Agile Coach is knowing and understanding the context within which the team members are working. The Agile Coach facilitates growth and improvement through empathetic understanding and influence rather than by just telling people what to do. Influence comes from building relationships rooted in respect, credibility, and trustworthiness.

In order for people to hold themselves accountable and each other accountable, they have to be comfortable with each other. People being comfortable speaking honestly about what they know or don't know and what they have or have not completed is fundamental to establishing the trust needed for accountability. Accountability and open communication are stepping stones for the resilience needed during this pandemic.

Something being taught this year to my second grade son is a fixed mindset versus a growth mindset. The fixed mindset is based on fear. People with a fixed mindset often fear failure, fear looking stupid and do not want others to see them as being wrong. A growth mindset is based on accepting losses, a desire to learn, being confident enough in one's abilities to be okay with not knowing, asking for help, making mistakes or not always being right.

Agile is very much about having and spreading the growth mindset, but this does require trust. The current company that I am an Agile Coach for values the concept of "believing those you work with are well intended in their interactions." I have found this statement alone to have a significant impact on company culture and trust.

Even though the workforce is largely isolated at home, they must still hold themselves accountable for their work responsibilities. Workers may be worried about spouses or themselves losing their jobs and the change to the family economics. Losing their homes which is bad enough now also means losing or moving their offices and classrooms since these functions are combined in their home.

These concerns are formidable, but employers must still be able to expect productivity will continue despite all the worldly stressors. This is why creating a culture that fosters trust and accountability is so important.

May 2020, only a couple of months into working remotely in Minneapolis, MN, the city went through an event of historic proportions with racial unrest triggered by the death of a man known as George Floyd. His death sparked protests and riots around the world. In Minneapolis and St. Paul, businesses were burned to the ground for miles and miles.

For the first time in my lifetime, I watched live on YouTube as the Minneapolis Police Department 3rd precinct office was taken over by rioters and started on fire. Many of my new team members resided in South Minneapolis within blocks or miles of where George Floyd died and where riots continued uncontrolled for days to weeks.

With thousands upon thousands of protesters and the National Guard being brought in, my husband and I sat with our three kids. We could hear chanting, protesting, banging, honking, screaming, etc. all night long from our home and more fire truck sirens than I've ever heard in my life.

Protesters, police, business owners and residents were injured, and several casualties took place during these racially fueled riots. Most commercial buildings for miles in every direction remained boarded up, some due to lack of funds to repair. Other businesses chose to keep boards up in fear of potential unrest.

George Floyd was killed in 8 minutes and 46 seconds with an officer's knee on his neck repeatedly telling the officer, "I can't breathe". I am a little sister, wife and mother of loving, gentle black men and also the granddaughter of a noble and honorable Minneapolis police officer. Like many others in the workforce, I can't describe the emotional hurricane I was carrying. The whole thing was so traumatic and close to home in more ways than one.

Could I focus on work the next day like everything was okay? No, I was not as sharp.. Did I expect my teams to focus at work the next day as if nothing happened? No. At work the following day, rather than the usual work driven questions, I took a softer approach. I asked if anyone needed someone to talk to or if anyone needed to take time off to process the recent events.

Some may wonder how this could possibly be the right approach. Some may argue not to mix personal and professional life compartments, keep the focus on the business and the product. Employees are human first. Employee retention, product quality, and delivery will be better once the smoke clears. Give people time to process the recent double whammy, back to back historic events in the comfort of their employers' support. It is just flat out the right/ethical/moral thing to do at this point.

To encourage healing and show that support to employees, organizations encouraged employees to get involved in the community rebuild. Businesses took it upon themselves to help other businesses in need.

It was evident that small, local businesses who were already suffering through the pandemic financially, lost whatever was left in the riots/fires/looting. So how do we keep everyone's focus on work and delivery at this point? You don't... You just don't.

Once your people have done what they need to do for personal healing, they will be ready and eager to refocus their undivided attention on work. They will be delivering stronger than ever. By empowering employees to communicate openly and hold themselves accountable to get work done, the organization has helped develop a sense of resilience.

Use this traumatic unrest as an opportunity to look around and see what can be done to contribute to a more diverse, welcoming and fair environment. Start those conversations. I reached out to others in my organization to collaborate and look for ways to acknowledge and address biases and racial/gender imbalances in our workplace.

It is the diversity of individuals working together as a team that makes the products special and unique. Diversity is to be respected, encouraged, and understood. Foster diversity and be supportive of each other's strengths and weaknesses, as well as different situations during this time.

Your employees are living through a global pandemic, homeschooling children, wearing masks, racial unrest and crime spiking. They may struggle to focus. You need to be aware and willing to engage in ways that are helpful.

The question is how to acknowledge, help and move forward in this climate? It is well worth the time effort to find ways to increase trust with employees, increase transparency in communication and empower teams to be accountable, in order to build resilience in your work environment.

Chapter 4

My Experience Being Hired During the Pandemic

I was invited into the office of my prospective new employer for multiple days of interviews with pairs and groups of people. The office is located in the North Loop, Warehouse District of downtown Minneapolis, Minnesota. This area of downtown is bursting with artistic, modern and sleek vibes. I grew up just seven miles north of the office where I was interviewing. This location was an excellent fit for me.

I arrived at the designated address, a dark red brick building with royal steps and two welcome lights on each side of the double doors. Previously the office building was a parking garage that was renovated into a modern workspace with a coffee shop feel.

My prospective employer was given the honorable award of "Best Place to Work" by Minneapolis Business Journal, and they were also awarded the "Coolest Office" by Minneapolis Business Journal. The interview went very well. I was blown away by the innovative thinking, down to earth approach and the inclusiveness of my potential co-workers. The leadership's direction and vision were highly compatible with hiring an Agile Coach.

The Agile role in this organization is everything you would want in an urban tech start-up. I felt I could plant my roots and settle down. I finally found "the one". The company's mission to improve healthcare outcomes felt good to me.

March 9, 2020, I got the call I'M HIRED!!! March 16, 2020 is set to be my first day. Yessssssss!!! I just landed my dream job at my dream company. This is GREAT! I'm overwhelmed with excitement. What could go wrong?

March 15, Governor Walz issued a stay-at-home executive order. My first day at the new job was March 16. Based upon the executive order, effective immediately Monday, March 16 schools also moved to a new, unheard-of distance learning method of online education.

It was the first day at the new job and the first day of distance learning for my children. Working parents, what does this mean? Can we say stressful? In my scenario, I am the mother of 3, but I am also the primary financial provider for our family of 5.

As you would expect, on the first day of distance learning, no one knows what's going on. Parents struggled with how to access their kid's work or set up their learning paths; how to connect with the teacher and other students. Not to mention the struggle of figuring out the eight different apps that must be downloaded along with seven different websites. The stress for me was in the balancing act. I cannot fail at parenting. I also cannot fail at being their provider.

Going into my first day of work and the first day of distance learning at the same time was tremendously unsettling. I considered my own family lucky because we had internet access already set up and computer devices for our children to use. We had a two-parent household with one parent who could come home on unpaid FMLA (family medical leave of absence) to help the kids. These things were considered luxuries during the pandemic. Even with all those advantages, it still was unpredictable, unstable and a lot of pressure.

Monday morning, March 16, first day. I drove into an empty downtown to pick up my laptop. Downtown was a total ghost town with a grey sky and fog.

I parked in an empty parking ramp, listening to the echo of my car door close and not a person in sight. I walked down the steps out to the street, careful not to touch any of the railings or doorknobs. A flashback from the 1991 film *What About Bob* ran through my mind as I pulled a baby wipe out of my pocket to open the stairwell door that led out to the sidewalk. After about a block and a half walk, I went up the steps into my new empty office building.

Just a few weeks prior, this same place had left such a vibrant and profoundly energizing impression. Inside stood the Human Resource Generalist, IT specialist with my computer and my boss.

I was given my work computer, a rundown of everything I needed to know and sent home to get started. Due to the COVID-19 pandemic, my first day was also the organization's first day transitioning to working remotely in an effort to help slow the spread of the virus.

By 10 AM on the first day, I was headed back home, unsure what to think. My mind is running, and the world feels eerily quiet as I drive. I grabbed a cup of coffee from the gas station on my way home. Usually, a cup of coffee is somewhat comforting. Except on March 16 walking into the gas station for coffee felt like a supply run on the T.V. series *Walking Dead.* Is this real life, right now? Everything is so weird, WTH?!?!! Back in the car, I turned the music off completely. Rather than running from my feelings, I chose to allow myself to feel all the disturbed emotions.

My mind was racing with questions. How will I meet everyone? How will I get included and socialized with my new co-workers? How will I build relationships across the organization and facilitate team building activities?

What if I miss meetings accidently or don't respond fast enough in the beginning while I'm settling into distance learning? This would be a terrible and unfair first impression.

One of the Agile principles mentions face-to-face interactions as the most effective communication method, how will I compensate? Another Agile principle says to make sure that everyone has the environment and support they need and then trust them to get the work done.

How will I assess their environment during a pandemic, and how will I give them the support they need? How will I thrive as an Agile Coach via 100% online work? What if I fail virtually right off the bat at my dream job and no one even has a chance to know me in person?

How will I help my teams adjust to their new setting when I don't even know what's going on myself? In the pandemic online environment, I have lost the ability to casually stop by desks and check on my teams throughout the day or ask a quick question. I won't have those break room conversations that build camaraderie. How will the computer set up work at home?

There are plenty of things to think about. My wheels are spinning a mile a minute my first day, and first few months, honestly. As with most things in life, you can worry yourself to death and it will do you absolutely no good at all. Roll up your sleeves and enjoy the adventure.

During a pandemic a new hire would need to balance the product deep dive with quickly learning about the new COVID-19 related impacts and changes needed. Learning COVID-19's impact and redirection was a top organizational and individual priority.

Chapter 5

Agile on the Job

Work Environment

"Ensure that everyone has the environment and support they need and then trust them to get the work done." – Agile Manifesto

How will I assess their environment during a pandemic, and how will I give them the support they need?

For financial support our HR generously offered to reimburse people up to a predetermined amount, if they did not have essential equipment/environmental needs met to work from home.

Whether the HR did or didn't assess and offer supportive solutions for remote workers, an Agile Coach should survey their teams. This can be merely checking in privately with folks on your team and asking them if they have what they need or if they're feeling stressed in their environment.

Traditionally, an Agile Coach may casually stop by desks to see how people are doing and make sure they have everything they need to get their work done. In a remote working environment, such check-ins need to be more intentional and deliberate. . During a pandemic the Agile Coach needs to reach out individually more often and in team settings to find out how people are doing and to make sure needs are being met.

You never exactly know what you're going to get when you reach out about a work environment. You may hear everything's fine. You may hear someone needs an office chair. You may need to reassure a terrified parent with kids home that everything's okay, let them know their employer is behind them 100%. Do the best you can, and we're in this together. You may hear someone's puppy barking, a doorbell ring, a child crying, etc.. You never know. Just listen with your problem-solving hat on from one human to another.

Early on, myself and the Product Owner got together to work with the teams on creating a pandemic team charter. We established and clarified our teams working agreements during the COVID-19 pandemic. During the team charter discussion, we focused on setting expectations for how we are going to support each other and work together through the pandemic.

The theme of the team charter was human to human, we will do our best; you are human first, an employee second; we're in this together. Our teams went through a lot this year emotionally, mentally and physically.

The technical computer environment at home is time consuming and intimidating even with help desk support assisting with the initial set up. The key for everyone is to be flexible, understand, be human and attempt to solve an issue, independently before asking for help.

On the other hand, don't waste a ridiculous amount of time NOT asking for help either. Be mindful, be compassionate, when a team member can't get into the meeting due to technical difficulties or someone's wifi going down midday.

Remind yourself everyone's doing the best they can. Parents are trying to rearrange their houses to find work and school spaces for themselves and their children.

This is not business as usual, but it is business. The job is getting done to the best of our abilities. It is important to be flexible and accepting of each other in order to accomplish the common goals.

Many families decided to sell their houses and move (including me) during this pandemic to find a better fit for the situation, not knowing how long this will go on. Be mindful in the current workforce of the teammates who are trying to align their personal and work lives.

Trust the people you work with and the people you've hired that they are doing the best they can to bring their whole selves to work in creative and innovative ways, regardless of what wild jungles they are trudging through.

Building Relationships Remotely

Coming into a new job and immediately telling people what to change and how to change it before you've taken the time to understand current circumstances undermines credibility and trustworthiness. Providing critical feedback without building relationships tends to raise defensiveness, so whatever you were trying to help with is down the drain.

At an organizational level, relationships are also extremely important for getting work done. Without trusting cross departmental relationships, the Agile Coach may not receive the prioritized help they need to remove an impediment for their team. Building relationships is one of the essential first steps for this job after getting hired.

Face to Face Interactions

"The most efficient and effective method of conveying information to and within a development team is face-to-face conversation." – **Agile Manifesto**

During the pandemic where in person face to face interactions are not possible, turn the video on for every single call. Encourage video engagement by setting an example, without forcing your team and others in the organization to do the same thing.

Video Engagement

Have some fun with creativity and encouragement. Make people feel comfortable. I have kids. I have a dog. I haven't had *Chip and Joanna Gaines* in here to fix up the house, professional lighting or a high-quality webcam. It's not about any of that. It's about coming as you are, wherever you are, and connecting with the team on a genuine personal level.

Interestingly, I found that the remote videoing made me feel closer to the teams than I would have felt at work due to having the video on. You have a unique opportunity to get to know people in their homes (literally) and in some cases, you have the honor of meeting their loved ones who are home with them, kids, partners, pets, etc. Experiment and get creative with face to face interactions to influence wide-ranged participation.

What has worked well for myself and other Agile Coaches I've talked to is creating sprinkles of fun. I do "would you rather" Monday's with zoom background themes and Friday's "get to know you" question themes.

This sounds mushy, but it does keep it interesting, builds relationships and encourages engagement. Some ideas are crazy hat day, pajama day, or video background themes, such as changing your background to a photo of anywhere in the world you would like to travel.

Video activities encourage the team to participate in turning the video on without twisting anyone's arm. Depending upon the current state of the organization's culture, regular video engagement could be phased in by starting with once a week video activity. Build up to making this a normal way of working, over the course of time.

One on Ones

I kept these one-on-ones short and sweet. Prepare questions and topics ahead of time and include the conversation points in the invite. Additionally, while getting to know people I would keep handy some questions and topics of interest on a notepad near my laptop for things to continue talking about if the conversation went dry on video.

When building relationships, be sensitive to people's limited time and busy schedules. Do not go beyond the meeting time. Check with them at the end of the first meeting on their preferences for recurring meetings to keep the relationship growing in a way that works for them.

I would also turn on my video camera during these meetings. If others didn't turn on their camera, I still kept mine on. I felt it gave a personal impression and connection for the person on the other end. One of the most important jobs of an Agile Coach is to build relationships.

Team Building

How will I get included and socialized with my new co-workers? The Human Resources Director did an excellent job facilitating many organization-wide activities that provided ample opportunity to socialize. These included shows and tell, yoga, virtual bingo, virtual lunch, virtual happy hour, etc.

While HR set these activities up at my current employer, it is important to collaborate with your boss and HR regarding organizational team building. Feel free to take the lead on organizational team building if these activities aren't already established. It builds culture and encourages socialization—influencing bonding without forcing it.

Extroverts might be struggling with working from home. Introverts may love it. Regardless of which side of that fence you're on, you may hear complaints such as these socialization opportunities are corny or annoying.

Don't let negative complaints deter you. Keep the socialization optional, not mandatory, and know that you're never going to have 100% approval of anything you do, ever, anywhere in life. Just trust the process. Soft techniques have been proven effective in team building. Team building activities make such a difference.

How will I build relationships across the organization and facilitate team building activities? I continuously experimented with what worked for various teams' different personalities.

A favorite team building activity turned out to be Jack Box, a virtual game that can be downloaded online for around $25 and comes with a pack of virtual games for groups. We did this team building activity with a portion of our retrospective event time.

Other idea's I experimented with were classic icebreaker questions (easily found online) to start or end team events. Another fun activity is adding music to virtual team events such as themed music playing when teams join the meeting.

Developing team building events is the fun part. You can be as creative as you want. You don't want to annoy teams, but you do want to experiment, find the balance of suggestions and trials without detracting from the team's soulful purpose.

Satisfying Customers with Continuous Delivery

"Our highest priority is to satisfy the customer through early and continuous delivery of valuable software." –
Agile Manifesto

Let's break that down into two parts, satisfying the customer and early continuous delivery of valuable software. In my case, starting a new job remotely during the pandemic I first had to go digging. In normal times, product research would only be a small portion of the new hire learning, since there would be natural, casual conversations taking place in the office that would help someone get up to speed.

In addition to reading documentation, I set up time with other Agile Coaches, the Chief Architect and Marketing Manager to learn about how the product operates and learn about where the organization would like to drive the product.

While I was trying to learn about this new product during the pandemic, the highest priority for learning was emerging from the Data Analytics team. Organizations during a pandemic don't need or want the usual product. Organizations want to know how the pandemic is affecting the customer's behavior and interests relative to their business model. The Data Analytics team quickly provided answers.

They studied and quantified consumer needs and interests for the clients. Understanding first what customers need, want and what they currently have is the basis for helping the customer no matter what in the world is going on.

Satisfying the customer during a pandemic is different than business as usual. In normal times a new hire would deep dive into the product to learn how to satisfy the customer.

Now for the second part, early and continuous delivery of valuable software during a pandemic is more challenging and important than ever. An Agile organization can adapt quicker to frequent deliveries of working software. Seeing how the organization is doing on continuous delivery metrics is helpful.

High Level Metrics Do Help

My organization already had valuable metric tracking in place for the Kanban teams when I was hired. Being hired in the pandemic, it is even more important to quickly learn what the metrics are, understand how they are being used and the data source.

The urgency in determining whether or not the delivery processes are working cannot wait during a pandemic. Businesses are closing rapidly based on dynamic market conditions. During normal times organizations have had more leeway to respond when something is not going well.

In order to ensure continuous delivery of working software, I reached out to the Agile Coaches who established and worked the metrics. Additionally, for hands on learning, I tinkered around in Jira's issues/filters section, gathering data and seeing if I could dig deeper to uncover more team patterns beneath the surface of metric total numbers.

These metrics are useful for early and continuous delivery to ensure they are on track and/or getting better and better over time. These numbers could be used to monitor and make sure the Agile teams stay on track for continuous delivery equal to or better than before the pandemic.

Suppose the organization did not previously have any metrics to monitor continuous delivery. I might suggest creating some metrics that are high level, yet not too detailed. This will help you prove your success working remotely. Honestly, leadership will likely sense if the company is thriving or failing anyway, but this is still an excellent way to put some data behind it and give recognition to the teams for their hard work and success.

The goal is the opposite of micromanagement with detailed metrics. High level metrics fit better with Agile values by trusting the teams to get their work done and allowing teams more flexibility. High level metrics will still give leadership the insight they need to guide the teams in the right direction.

"Continuous attention to technical excellence and good design." - Agile Manifesto

"The best architectures, requirements, and designs emerge from self-organizing teams." - Agile Manifesto

Less is more for metrics. Good metrics demonstrate if the features are on track for continuous delivery, the quality of software being delivered, and if the customers are satisfied. For Scrum metrics consider including velocity of the team. For Kanban teams consider including things like throughput and cycle time.

"Agile processes promote sustainable development. The sponsors, developers, and users should be able to maintain a constant pace indefinitely." – Agile Manifesto

Some additional helpful metrics could be related to operations focus percentage, strategic focus percentage, and bug percentages (quality assurance measure of bugs/defects). Bug percentages are significant for making sure that working software is the top goal. Having some form of simple, high-level metrics helps an organization keep an eye on frequent delivery and quality.

"Working software is the primary measure of progress." –
Agile Manifesto

Adjusting to Change

Adjusting to changing priorities is usually in the product team's court to call the shots on, however everyone has a responsibility to do their part in supporting this process. Keeping up with changing requirements means working with the Product Owner to make sure the team's backlog is always up to date and in priority order.

It is important to ensure that the team has what they need to get the work done as soon as the need for change is noticed. Another form of support is reminding the team of their goal - to satisfy the customer. The ability to embrace change is the skill that builds product resilience.

Business and Development Work Together

Finding a way to connect your development team to the end user is controversial but can be rewarding with Product Owner involvement. In person meet and greet, question & answers sessions can be enlightening and motivating. Some people find it easier to keep the end user in mind after meeting them and learning about their needs in real life. Having a face and name in mind for the work being done can make a big difference.

Traditionally, organizations collect feedback through surveys and pass these results on to Product Owners and Development Teams. This is a valuable method, too.

"Business people and developers must work together daily throughout the project." - Agile Manifesto

Scrum Masters and Product Owners can be thought of as a married couple; parents for the team. The Product Owner is the business representative on the team. The Product Owner provides the vision and direction needed for the team to get their work done.

The Scrum Master helps facilitate discussions between the team and the Product Owner, as well as escalating anything standing in the way of progress.

During the pandemic without casual office conversations this may mean specifically reaching out to your Product Owner to see how they are doing. Are they overwhelmed? Do they have the bandwidth to support the team? Are they getting what they need from the business side in order for them to guide the team?

It may mean reaching out to the developers, too, for a pulse check. Ask developers (1.) if their priorities are clear and concise (2.) do they know what to work on next and (3.) do they have the information they need from the Product Owner to do the work.

Retrospectives

"At regular intervals, the team reflects on how to become more effective, then tunes and adjusts its behavior accordingly." - Agile Manifesto

Retrospectives are great for the team. Reverse retrospectives are a good opportunity for the team to openly tell the Scrum Master and Product Owner what is and isn't working. The Scum Master and Product Owner can use this opportunity to identify ways to improve their servant leadership for the team.

Explore the idea with the team of doing the reverse retrospective anonymously. Some teams will like this idea to offer suggestions anonymously, while others will prefer to give direct feedback and have an open dialog about their opinions.

There are anonymous retrospective tools, easily found online. They are fantastic, especially if you are new, working remote and/or the team is not comfortable providing direct feedback. As an Agile Coach, part of the role is guiding and coaching across the organization including product and business departments.

Helping them iron out any issues they have going on will essentially come back full circle to directly helping the team. Ultimately, Product leadership should make sure there is a clear vision and clear priorities for the Product Owners to guide the team with.

Agile Mishaps

You might have heard things like "Agile is Dead", and "it does not work". Or you might have read a blog that says "Agile is outdated, and awful". I beg to differ, but I will say that it will be problematic rather than beneficial if misused or implemented improperly.

For a robust Agile culture, invest in and establish a strong foundation, initially. Keep a team around to maintain the momentum of continuous improvement, even after the initial implementation.

This will ensure continued success of the initial investment. Without these careful steps, you may well wish Agile would disappear. A self-proclaiming Agile organization that is not actually Agile will also make progress harder for the next Agile Coach or Scrum Master.

The Agile Coach/Scrum Master will be tasked with fixing all the misconceptions for a developer or leader who despises what they "think" Agile is. Agile will not work effectively, if it is not implemented and used correctly. User error accounts for a majority of the bad reputation.

I believe the principles of Agile can be applied anywhere in business or even personal lives. However, there are circumstances where Agile would not be the best fit. For example, a client who doesn't have the time or is not interested in being a frequent participant in the short feedback loops. This is rare, but it does happen for various reasons.

People who have experienced Agile mishaps may be more likely to say Agile is dead. The poor outcome could be related to having an over stretched Agile Coach/Scrum Master or Product Owner, rather than Agile itself being the issue. I've even seen companies try to combine these roles into one person, a Scrum Master/Product Owner.

Scrum Master and Product Owner roles require two full time jobs. Combining these two roles creates a conflict of interest on where they should be focusing their time and attention. Not a good idea - teams, clients, stakeholders all need to put some sturdy struggling boots on if the Scrum Master and Product Owner roles are combined.

Understanding the structure of an Agile team also ties into the budget to some degree. Make sure the organization can afford to have the full teams that are needed to thrive. Having an underfunded budget for the Agile transition will hinder the ability to implement this reliability from the beginning.

My advice would be to evaluate organization size and teams before transitioning to Agile. Get a ballpark financial estimate including training and hiring costs beforehand. Then start where you can, and in true Agile fashion, deliver it incrementally.

Start with leadership training across the board. An important financial and strategic consideration is staffing Agile teams appropriately. In the past there was one Project Manager for many projects i.e. five-fifteen projects. With Agile, there would be no Project Managers, but instead one Scrum Master and one Product Owner per two teams.

While the initial investment in hiring enough of the right people is substantial, the results will be equally substantial. Once the hiring is complete those teams will function far more efficiently by having dedicated people closer to the team to escalate issues, remove roadblocks quicker, improve processes for maximum delivery.

For terminology sake, when implementing Agile, there technically shouldn't be any more "projects". The terminology "projects" is replaced by the Agile concept of "features" or "backlog features". An Agile team should work cross-functionally, ideally working without strictly defined roles to deliver a series of related features. The terminology and reality are both different. Agile deals in bite size deliverable (features). Whereas, the word project is associated with one deliverable at the end.

Another common Agile mishap that might lead someone to feel Agile is dead is simple role confusion! In Scrum, for example, role confusion becomes a considerable roadblock to Agile success and can leave people with a sour taste left in their mouths. Where's Abbott and Castello? "Who's on first!?!?!?" Are you prioritizing the backlog, or am I? I thought you were doing it. But that's your role! See where this is going?

For the record, Product Owners own the backlog and the prioritization in Scrum. The Product Owner is the primary business stakeholder for the team as far as providing the vision and prioritizing the work coming from the business or client. The Scrum Master has a crucial role in facilitating the conversations, helping identify red flags and remove impediments early on, and providing coaching where appropriate. When these roles are confused, the team will be confused. Role confusion comes from a lack of synchronized training.

In a pure, highly advanced Agile environment, the teams should be fully cross-functional. Anyone can do anything to achieve the team goal - because one team, one goal. However, that's not always possible, due to lack of training, knowledge etc., but it remains a desirable target for Agile teams. A few Agile mishaps does not mean Agile is dead.

Chapter 6

Conclusion

I intentionally chose not to go into too much detail regarding the various frameworks and their structures. There are many resources about framework specifics. At this time, I wanted to keep the focus on Agile at its core. What the principles are and what the values are. 2020 has been a heck of a year.

Even the most Agile people and organizations have struggled. Celebrate your accomplishments as big or small as they may be. Celebrate diversity in your organizations and individuals. Be kind to your neighbors. You made it through the year. You learned a lot. You enhanced your Agility by leaps and bounds. May 2021 be ever in your favor. Be Agile world - until next time.

-Christina S.

www.ingramcontent.com/pod-product-compliance
Lightning Source LLC
Chambersburg PA
CBHW070452220526
45466CB00004B/1803